7692

Rourke
Educational Media
rourkeeducationalmedia.com

Animals Have Classes Too!

Arthropods

Before, During, and After Reading Activities

Before Reading: Building Background Knowledge and Academic Vocabulary

"Before Reading" strategies activate prior knowledge and set a purpose for reading. Before reading a book, it is important to tap into what your child or students already know about the topic. This will help them develop their vocabulary and increase their reading comprehension.

Questions and activities to build background knowledge:
1. *Look at the cover of the book. What will this book be about?*
2. *What do you already know about the topic?*
3. *Let's study the Table of Contents. What will you learn about in the book's chapters?*
4. *What would you like to learn about this topic? Do you think you might learn about it from this book? Why or why not?*

Building Academic Vocabulary

Building academic vocabulary is critical to understanding subject content.
Assist your child or students to gain meaning of the following vocabulary words.
Content Area Vocabulary
Read the list. What do these words mean?

- *abdomen*
- *combined*
- *jointed legs*
- *microscope*
- *molt*
- *segmented*
- *thorax*
- *trunk*

During Reading: Writing Component

"During Reading" strategies help to make connections, monitor understanding, generate questions, and stay focused.
1. *While reading, write in your reading journal any questions you have or anything you do not understand.*
2. *After completing each chapter, write a summary of the chapter in your reading journal.*
3. *While reading, make connections with the text and write them in your reading journal.*
   a) *Text to Self – What does this remind me of in my life? What were my feelings when I read this?*
   b) *Text to Text – What does this remind me of in another book I've read? How is this different from other books I've read?*
   c) *Text to World – What does this remind me of in the real world? Have I heard about this before? (News, current events, school, etc....)*

After Reading: Comprehension and Extension Activity

"After Reading" strategies provide an opportunity to summarize, question, reflect, discuss, and respond to text. After reading the book, work on the following questions with your child or students to check their level of reading comprehension and content mastery.
1. *How do scientists classify arthropods? (Summarize)*
2. *What can you conclude from reading about myriapods? (Infer)*
3. *What are the four main classes of arthropods? (Asking Questions)*
4. *If you had to make a new way to classify arthropods, how many classes would you use? Why? (Text to Self Connection)*

Extension Activity
Find information about the tools scientists use to help them classify arthropods. Draw the one you think is most important. Explain your choice.

Table of Contents

Red cliff crab

Let's Classify!

Scientists classify all the living things on Earth. They look at ways living things are alike and different. They sort them into groups to better understand them. Kingdoms are the largest groups. Plants make up one kingdom. Animals make up another kingdom.

Animal Groups

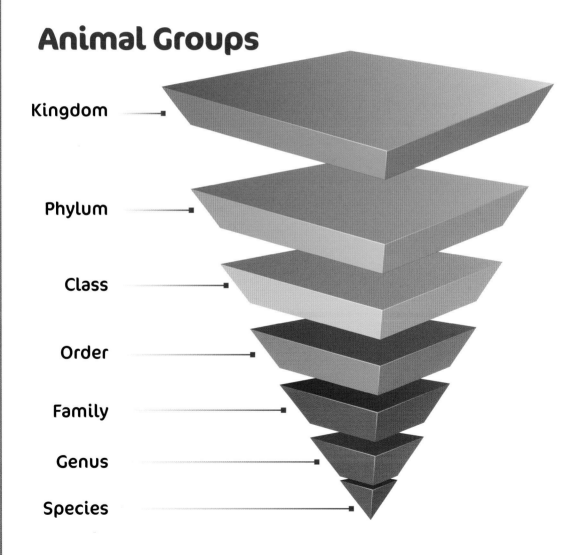

Kingdom

Phylum

Class

Order

Family

Genus

Species

The things that make up a kingdom are different in more ways than they are alike. Scientists break kingdoms into smaller groups. Each group is a phylum. Arthropods are the largest phylum in the animal kingdom.

There are millions of kinds of arthropods. They have **segmented** bodies and **jointed legs**.

Arthropods have no backbone. They have no bones at all! They have an exoskeleton. It is a hard covering. It helps protect the arthropod's insides and give it shape.

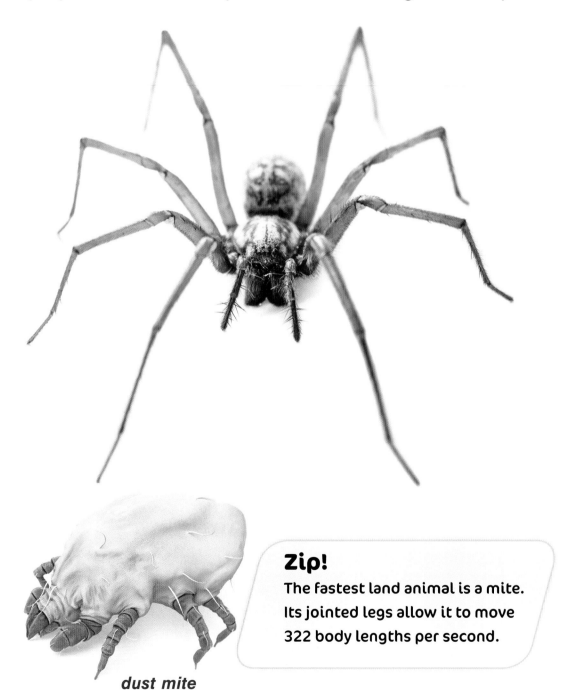

dust mite

Zip!
The fastest land animal is a mite. Its jointed legs allow it to move 322 body lengths per second.

Arthropod Classes

Scientists break phyla into groups called classes. Many scientists sort arthropods into four main classes. They are insects, arachnids, crustaceans, and myriapods.

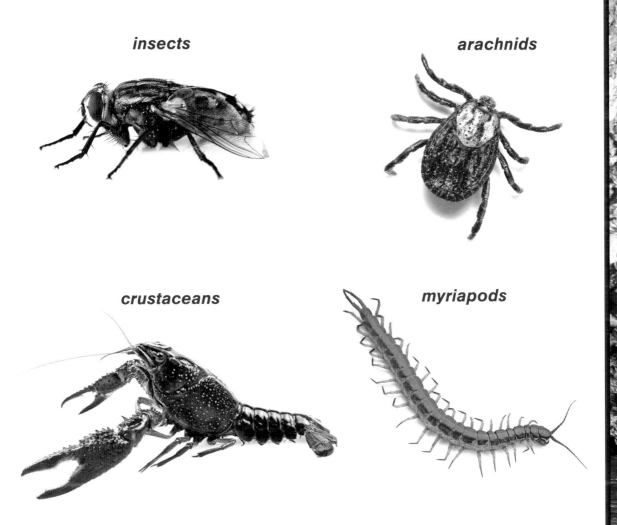

insects

arachnids

crustaceans

myriapods

Flies, bees, and butterflies are kinds of insects. An insect's body has three main parts: the head, **thorax**, and **abdomen**. Insects have one pair of antennae or feelers on their head. Most insects have six jointed legs.

thorax

head

abdomen

antennae

jointed legs

Sniff!
An insect's antennae do more than feel. Those feelers can smell too!

Arachnids have their own class. Each member has eight legs. It has two main body parts. Its head and thorax are **combined** into one part. Its abdomen is its second main body part. Arachnids have no antennae.

Spiders and scorpions are arachnids. Mites and ticks are also arachnids. Horseshoe and king crabs are in this class too.

Mexican red knee tarantula

crab

Crustaceans have ten legs. Each leg has two parts. They have two pairs of antennae. Some crustaceans have three main body parts like insects. Others have two main body parts like arachnids. Lobsters, crabs, and shrimp are in this class. Barnacles and woodlice are crustaceans too.

woodlouse

Long Legs!
Japanese spider crabs are the largest crustaceans. They measure up to 13 feet (4 meters) across with their legs stretched out.

Crustaceans also have a crusty or hard covering. They cannot grow inside their hard shells. They must **molt** to grow. The arthropod sheds its hard shell. A thin, soft shell takes the place of the hard shell. The animal grows for a short time. Then the shell gets hard again. This happens many times over its growing years.

horseshoe crab molted exoskeleton

Myriapods are the smallest class of arthropods. Centipedes and millipedes are myriapods. They have more than ten legs. They have two antennae. Their bodies have two main parts: the head and **trunk**. The trunk can have many segments.

Myriapods live on land. Some are too small to see without a **microscope**. The biggest are almost one foot (30 centimeters) long.

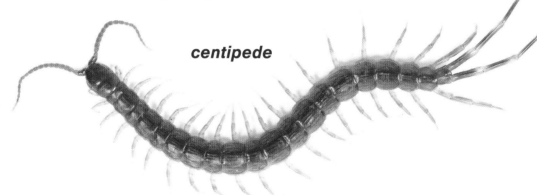
centipede

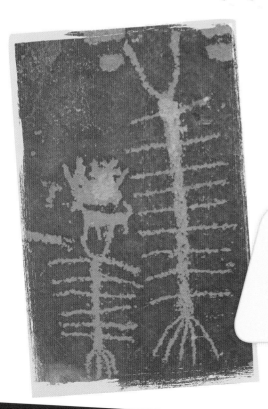

How Old?
Myriapods are among the oldest land animals. Fossils show they crawled Earth about 200 million years before there were dinosaurs.

Order, Please!

scorpion

Scientists break classes into orders. Scorpions and ticks are in the same class, but in different orders. There are 11 orders in the arachnid class.

cricket

Families are the smaller groups that make up orders. The members of a group are more alike as the groups get smaller. Crickets and grasshoppers are in the same class and order. They belong to different families. Crickets and grasshoppers are not closely related.

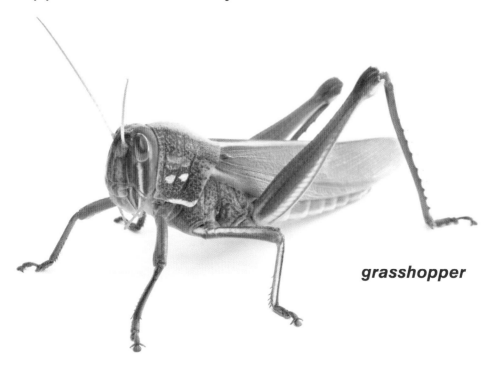

grasshopper

Families are also broken into smaller groups. Each group is called a genus. Members of a genus are closely related. The bluegrass crayfish and the hairy crayfish belong to the same genus. They are alike in many ways.

So Many Species

Each genus contains one or more species. A species is a single kind of living thing. Each species belongs to only one genus.

Spider!
The Giant Golden Orb Weaver belongs to the Nephila genus. This bat-eating spider can grow bigger than a person's hand.

A species has a common name and a scientific name. The scientific name has two parts. The first part is the genus to which it belongs. It always begins with an uppercase letter. The second part is the species name. It begins with a lowercase letter.

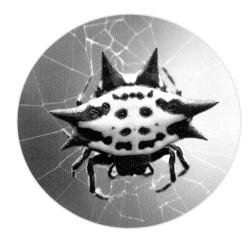

Housefly

Kingdom:	Animalia
Phylum:	Arthropoda
Class:	Insecta
Order:	Diptera
Family:	Muscidae
Genus:	*Musca*
Species:	*Musca domestica*

Spiny orb weaver

Kingdom:	Animalia
Phylum:	Arthropoda
Class:	Arachnida
Order:	Araneae
Family:	Araneidae
Genus:	*Gasteracantha*
Species:	*Gasteracantha cancriformis*

The *Coccinella septempunctata* is a species of insect. It is a kind of beetle. Its name shows it belongs to the *Coccinella* genus. The scientific name is the same all around the world.

Lots of Ladybugs

There are about 5,000 species of ladybugs. The seven-spotted ladybug has a shiny red and black body.

This beetle lives in the United Kingdom. Most people there call it a ladybird. It also lives in the United States. Most people in the United States call it a ladybug. Ladybird and ladybug are two common names for the same insect.

seven-spotted ladybug / ladybird

Scientists do not all agree on how to sort arthropods. Some sort arthropods into eleven or more classes. Some look more at how species are alike. Others look more at how they are different from each other. They classify arthropods for the same reason though. Sorting arthropods into groups helps us better understand these living things.

ACTIVITY

Build and Classify

Make your own arthropod collection to share with others.

Supplies

paper	clay
pencil	scissors
paints, crayons, or	glue
markers	pipe cleaners
paper plates	small plastic beads

Directions

1. Pick a class: insect, crustacean, arachnid, or myriapod.
2. Draw a sketch of the arthropod you would like to make. How many legs should it have? How many body parts? How many antennae?
3. Use your materials to build a model of your arthropod.
4. Label your arthropod with its class.

Make more models of other arthropods. Sort them into classes. Share them with your friends.

Glossary

abdomen (AB-duh-muhn): the rear part of an arthropod's body

combined (kuhm-BIND): two or more things put together as one

jointed legs (JOIN-ted LEGz): legs where the parts are connected by softer parts or joints that can bend

microscope (MYE-kruh-skope): a tool that makes small things look bigger

molt (MOHLT): to lose an old shell or covering so a new one can grow

segmented (seg-MEN-ted): made up of sections or parts

thorax (THOR-aks): the middle part of an arthropod's body

trunk (TRUHNGK): main part of the body that follows the head

Index

Show What You Know

1. Why do scientists sort living things into groups?

2. What do all members of the arthropod phylum have in common?

3. What are the main differences between insects, arachnids, crustaceans, and myriapods?

4. What do the two parts of a scientific name stand for?

5. Why do crustaceans need to molt?

Further Reading

Berne, Emma Carlson, *Crustaceans*, Capstone Press, 2017.

Dell, Pamela, *Arachnids*, Capstone Press, 2017.

Romero, Libby, *Ultimate Explorer Field Guide: Insects: Find Adventure! Go Outside! Have Fun! Be a Backyard Insect Inspector!*, National Geographic Children's Books, 2017.

About the Author

Lisa Colozza Cocca has enjoyed reading and learning new things for as long as she can remember. She lives in New Jersey by the coast. You can learn more about Lisa and her work at www.lisacolozzacocca.com.

Meet The Author!
www.meetREMauthors.com

PHOTO CREDITS: Cover and Title Pg ©Uwe-Bergwitz; Pg 3 ©xeni4ka; Pg 4 ©lvcandy; Pg 5 ©Steve Debenport; Pg 6 ©Martina_L, animatedfunk, Pg 7 ©ABDESIGN, ©yothinpi, ©zhengzaishuruAntagain; Pg 8 ©Antagain, ©lamyai; Pg 9 ©Okea; Pg 10 ©Delpixart, ©DanielaAgius, ©Wiki; Pg 11 ©BobMcLeanLLC; Pg 12 ©ncognet0, ©TommyIX; Pg 13 ©johnaudrey, ©Gearstd; Pg 14 ©PetrP, ©ithinksky; Pg 15 ©Guenter A. Schuster; Pg 16 ©photographereddie; Pg 17 ©DirkRietschel, ©dossyl, ©Antagain; Pg 18 ©ConstantinCornelDenisVesely; Pg 19 ©Antagain, Pg 20 ©sidsnapper; Pg 22 ©macroworld

Edited by: Keli Sipperley
Cover and interior design by: Kathy Walsh

Library of Congress PCN Data

Arthropods / Lisa Colozza Cocca
(Animals Have Classes Too!)
ISBN 978-1-64369-063-6 (hard cover)
ISBN 978-1-64369-076-6 (soft cover)
ISBN 978-1-64369-210-4 (e-Book)
Library of Congress Control Number: 2018956024

Rourke Educational Media
Printed in the United States of America,
North Mankato, Minnesota